THE GRAND UNIFIED THEORY

NISHANTH MEHANATHAN

I dedicate this book to my family, friends, teachers, Shree Krishna, Shani Bhagawan and Vaishnodevi Ma

Contents

Acknowledgements

I would like to thank all my teachers, friends, colleagues and family members for the warm support they have extended to me while working on this book, these were the giants on whose shoulders I stood.

The search for a "Unified theory" has been a tale spanning 2 decades. I came across it in my search for answers to "world saving problems" like "100% efficiency solar cells" and other problems like that. I would like to thank all the people who I came across on this journey for their kindness, ideas, clues and support for that is what made this work possible. Finally, I would like to thank Lord Krishna who made all of this possible.

Introduction

In this book, we discuss how the science of language leads to the Grand Unified Theory.

The technical topics covered to develop the Grand Unified Theory:

1. What Is An Object?
2. Classification Of Words In a Language
3. "*Satta*" Or "Existence/Being" is The Highest Universal

The study of these technical topics surprisingly leads to the "Grand Unified Theory". The "Grand Unified Theory" here means a theory which unifies not only particles and forces but "Everything" as a manifestation of a single fundamental state. This work finds its root in Sanskrit Language Sciences as presented by the famous Sanskrit grammarian *Bhartrhari* and other grammarians of the Sanskrit language.

CHAPTER ONE

A Thing, the types of words and word meaning

An object is defined as that which is the substratum of the genus, differentia and action [1] as depicted in Figure 1.

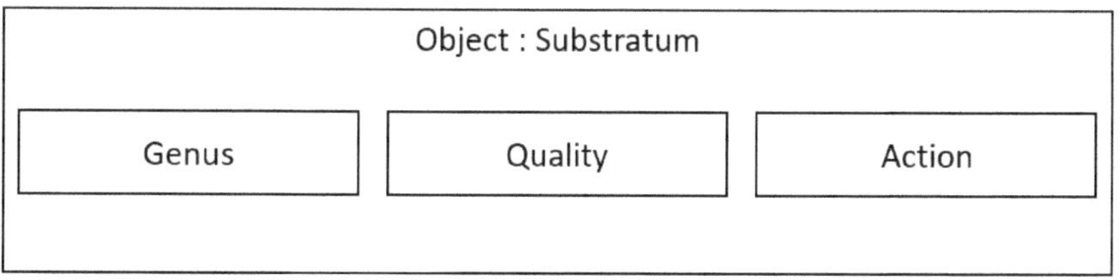

Figure1. Object as the substratum of genus, differentia and action

These attributes (Genus, Quality and Action) are classified as shown in Figure 2. An attribute is of two kinds- one that is inherent and the other which is imposed upon it like a name. The inherent attribute again is of two kinds, an attribute that is fully accomplished and that which is in the process of accomplishment. An accomplished attribute is of two kinds- the genus or class and the quality. Genus is never found away from the individuals in which it resides, while a quality distinguishes a thing from other things belonging to the same genus or class.

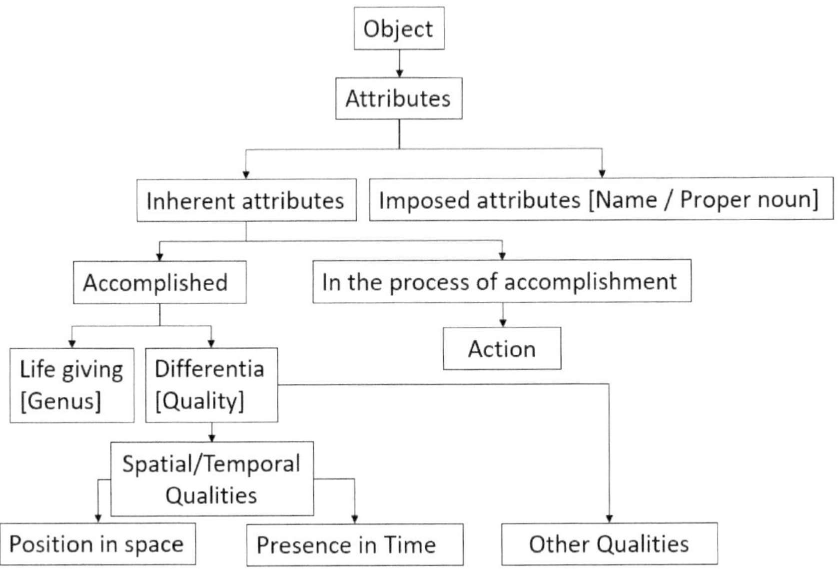

Figure 2. Object attributes, which uniquely identify an object

An example of the genus is fox-ness, quality is whiteness and action is cooking. An attribute of an object in the process of completion is called an action. For instance, the colour of a sheet of paper (whiteness) is an accomplished fact. But action implies a series of activities some completed and some in the process of completion, which occupies successive portions of time.[2].

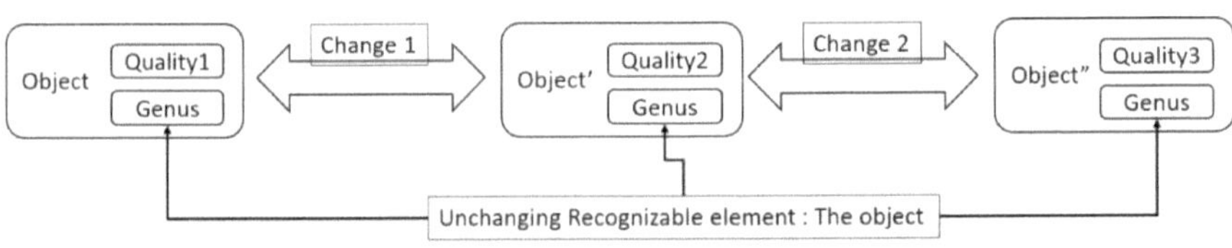

Figure 3. The unchanging recognizable element which remains across various changes

Object, its identity and the identity theorem:

Identity theorem: An object stands for that unchanging recognizable element which persists in all the changes that it undergoes [3] as shown in Figure 3. This is because if an object changes, it still remains the object hence the object is something which persists across the change and which we know to be the genus or universal.

For example, if a silver bell (or a silver object) is twisted out of shape what remains is it being made of silver and it is an object. So it is still a silver object but the silver bell is destroyed. So speaking practically it is the genus which the object is. Qualities are properties which undergo a change. As a result of the genus persisting you are able to identify the object across all changes as the genus remains constant.

Classification of words in the Sanskrit Language and the types of objects a word can refer to:

The Sanskrit linguists hold that the import of words is either genus, quality, action or names. Hence there are four main types of words- genus word, quality word, action word and name word [4]. It is particular to the Sanskrit language but the import or idea is universal to all languages. What is indicated by a word is of four types- genus, quality, action and proper names as depicted in Figure 4.

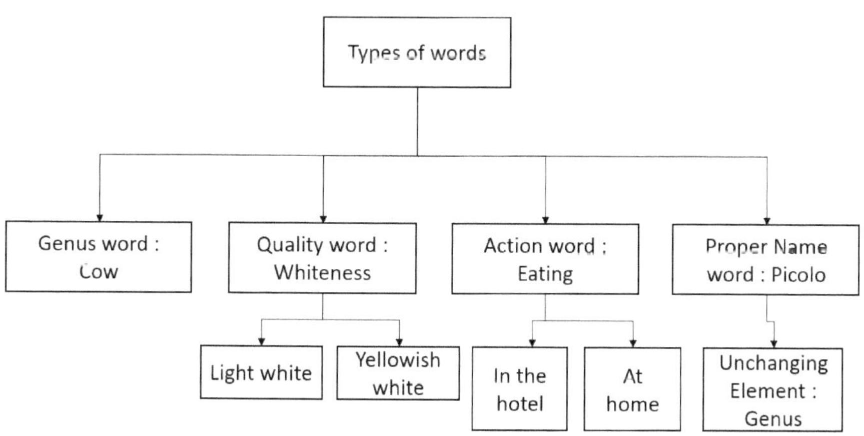

Figure 4. Types of words and what they indicate with examples.

Word Meaning is the Universal or Genus:

Theorem: Word meaning is the universal or what is indicated by a word is a universal.

This is the claim by the Jatipaksa school of thought and is also shared by the various grammarians.

Proof:

So as discussed before the four classes of things or words are:

1. Genus word.
2. Quality word.
3. Action word.
4. Proper name word.

Genus word: The universal/genus consists of all objects belonging to the class. Example: Tree. Genus: All trees.

Quality word: The universal consists in that they distinguish things from other things of a class. So the Genus consists of the things which they distinguish. Example: White. Genus word: All white things for example light white box, yellowish-white lamp. All these things produce the cognition of whiteness.

Action word: The universal/genus consists of all similar actions distinguished by subject, object, position, time etc. Example: Eat. Genus: All Eating actions are distinguished by qualifiers, space and time. All these actions produce the same cognition of eating.

Proper name word: It stands for that unchanging recognisable element that persists in all the changes it undergoes. This is because if an object changes, it still remains the object hence the object is something which persists across the change and which we know to be the genus. Hence the name also denotes the universal. Example: Picolo [4]. The four types of words have been listed in Figure 4. With their respective examples.

So we see that the four classes of words convey a genus only, a genus word already conveys a genus. A quality word conveys a class of qualities which fall within the same class for example dark yellow, light yellow etc constitute the class of colours called "yellow" which in itself is a quality. Action words too convey a genus of actions like "eating in a park", and "eating with hand" are all actions which come under the genus of eating (action). A proper name stands for that unchanging recognisable element that persists in all the changes it undergoes; which is by definition the genus. **Hence we can generalize that all words that indicate an object (genus, quality, action or class instance) convey the genus only.**

Notes

1. Charudev Shastri, Vyakaran Mahabhashya (Pratham Aahrikantri) – Page 3.
2. Sahitya Darpan, II.4, page 43.
3. Vakyapadiyam of Bhartrhari, III.1.11, III.1.12 commentary.
4. Makkhanalāla Śarmā, Bhāratīya kāvyaśāstra ke siddhanta - Page 69.

CHAPTER TWO

All Words Convey "Existence / Being"

As discussed a thing has four attributes:

1. Name
2. Class
3. Quality
4. Action

A thing and its various attributes are depicted in Figure 1 [1].

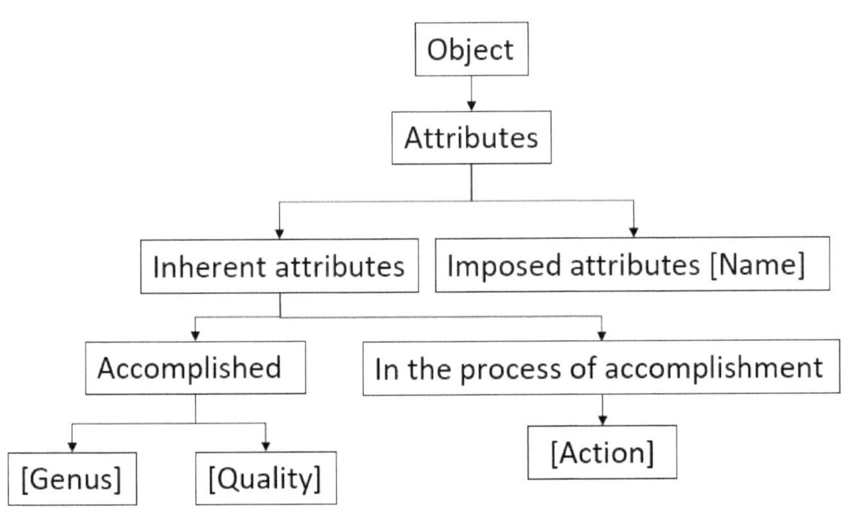

Figure 1. A thing and its attributes

Existence makes things exist, hence it is in everything: Existence means the "state of existing". Therefore, it is the class under which all fall. The highest universal is "Being" or "Existence" (Satta), and it is inherent in substances, qualities and activities and makes them exist [2]. Satta is all-pervasive and inheres in everything(that exists) and makes them exist, including substances, qualities, and actions. So everything falls under the category of existence(Satta). All words ultimately refer to Satta.

Word meaning, where word indicates a thing according to the two schools of thought in India is as follows:

1. Jatipaksa claim that the word refers to the class(Jati) to which the individual/thing belongs. What is implied is that the word refers to a persistent thing across changes(Jati) [3].
2. Vyaktipaksa claim that the word refers to the individual, which is a finished thing or an instance of a class(Jati) [3].

A word can mean either the individual thing which it signifies (finished thing) or the genus the thing has, according to the two schools of thought as shown in Figure 2.

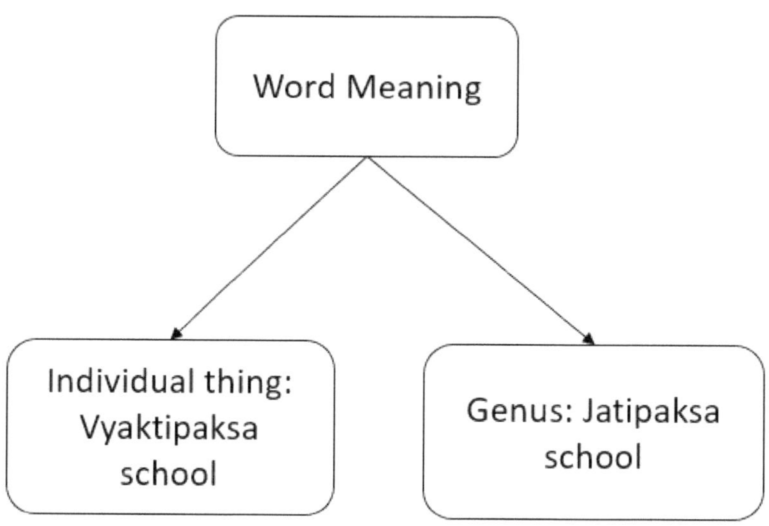

Figure 2. The two schools of thought and word meaning

The Individual or finished thing(Dravya) has the following properties [4]:

1. It is a collection of qualities, both generic and specific.
2. It has an essence(Tattva) which is not lost as qualities appear and qualities disappear. It remains the same even if some qualities appear and some qualities disappear. For example, an orange fruit at one instant is raw and in the future, it ripens, instead of naming it a different orange we call it the same orange as the essence is not lost only the quality of being raw is lost and the quality of being ripe gained.
3. The essence is the basis for the cognition of the same form and the naming of the individual and is a property present in the object.

Jati and the individual thing:

Jati is that attribute which gives life to a thing and is never found dissociated from the individual thing in which it resides.

The relation between the word and the denoted object must be permanent [5]:

The relation between the word and the denoted object must be permanent, if it is not so then the word would not refer to the object it was referring to initially an example is shown in Figure 3. The word must point to such a thing in the object which is permanent such that the relation between the word and the denoted object is permanent as shown in Figure 4.

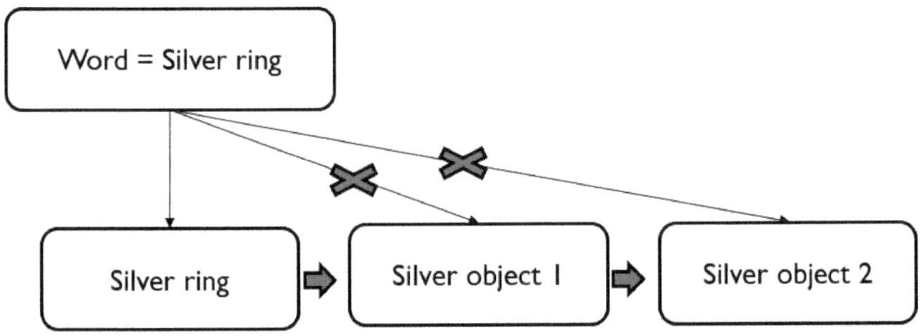

Figure 3. Word(Silver ring) stops referring to the denoted object after the silver ring becomes silver object 1 which is undesirable

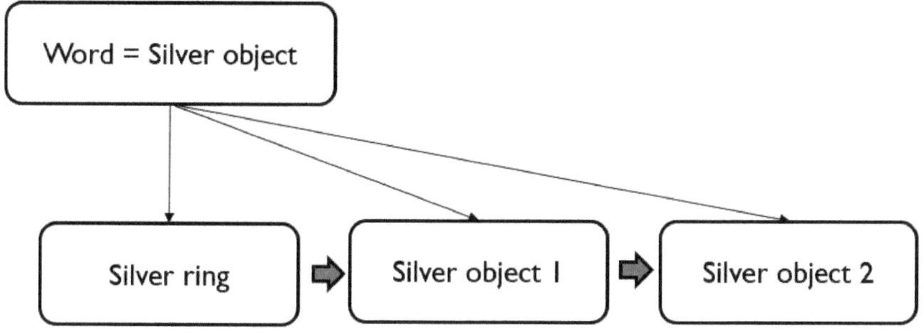

Figure 4. Word(Silver object) continues referring to the denoted object even after the silver ring becomes silver object 1 and silver object 2 which is highly desirable.

Persistent Jati, Jatipaksa school and persistent Dravya, the Vyaktipaksa school:

Case 1: When word meaning is a Jati: Jatipaksa school

If the word meaning is a Jati, then the highest Jati being Existence/Being or "Satta" can be taken as the word meaning this is because the highest Jati persists or remains with the object in all cases. An example is shown in Figure 5. As a result, the word continues to denote the object in all situations. So if the word meaning is a Jati, the words always refer to "Existence". And each object is a type of "Existence" differentiated by some quality or the other.

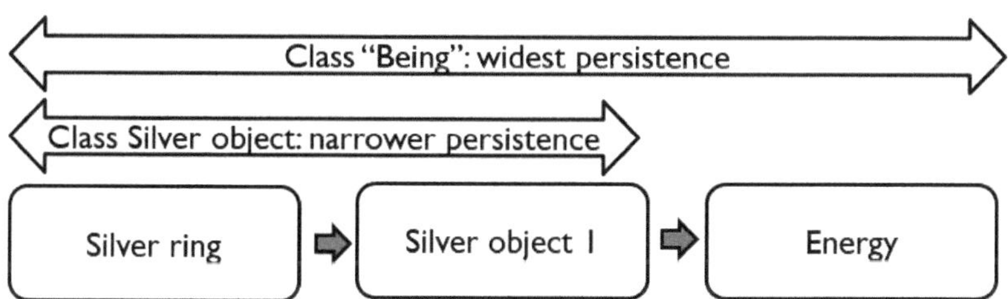

Figure 5. The class silver object partially persists, but the class "Being" persists across its existence, hence if the word was to mean "Being" it would be always associated with the denoted object.

Case 2: Word meaning is an individual thing or Dravya: Vyaktipaksa school

If the word meaning is an individual thing and the property which names that individual thing or "Dravya" is permanent it continues to denote that thing persistently. In Figure 6, we revisit the example tackled in case 1. If the naming property is "being existent" and the thing is consequently called "Being" or "Existence" the word would always signify the denoted object as shown in Figure 6. Dravya "Being" is the expressed meaning of all words or all objects are the dravya "Being" differentiated by some property or the other.

Hence the message is that both the meanings i.e. Jati and Dravya(Individual object) point to or refer to the same meaning "Existence" or "Being". Hence all words refer to or mean "Being"/"Existence" which is called *Brahman* (something which is present in all and is present everywhere) as shown in Figure 7.

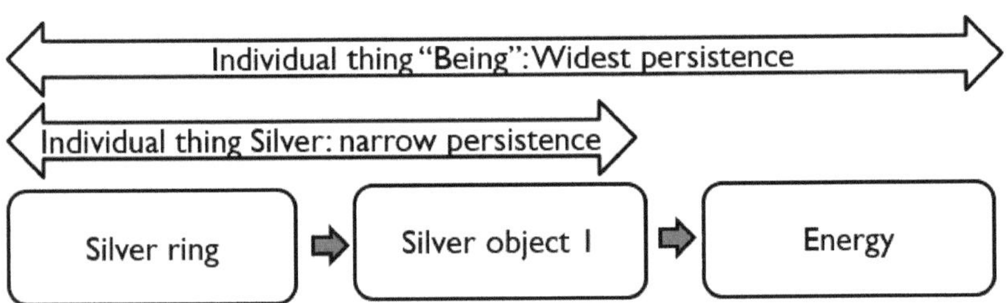

Figure 6. The individual thing Silver partially persists, but the individual thing "Being" persists always, hence if the word is "Being" it would be always associated with the denoted object. Where the naming property is "Existing".

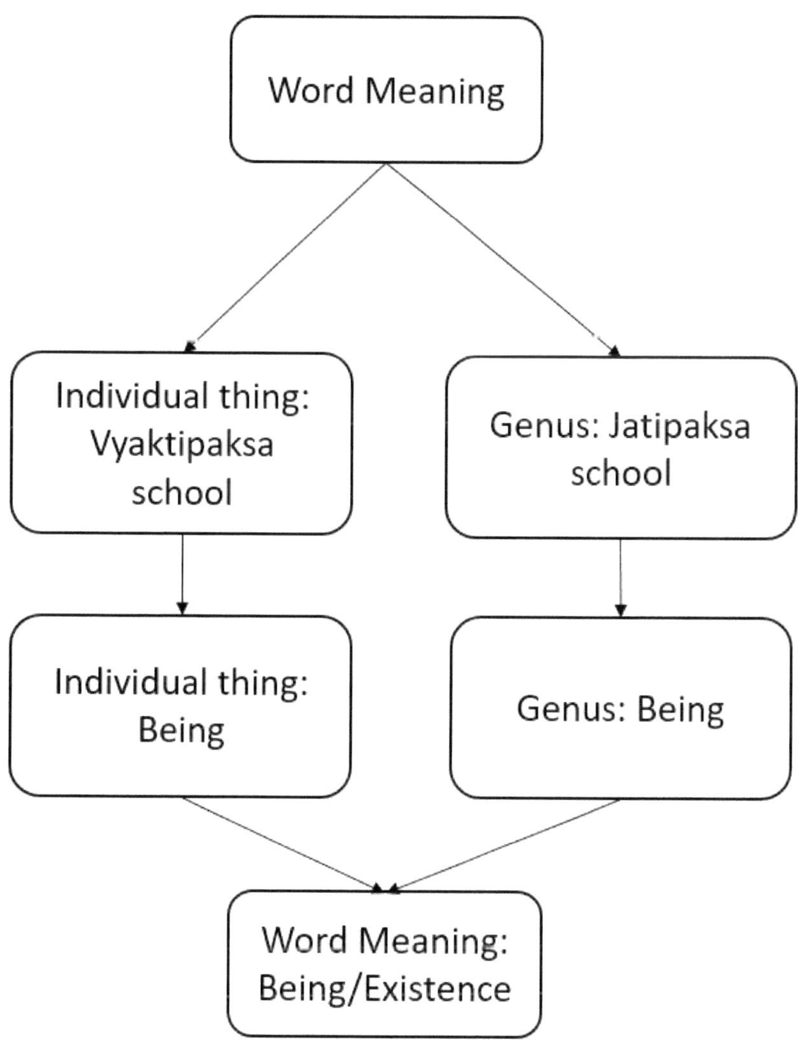

Figure 7. The two schools of thought on Word Meaning and Existence

Being or Existence is of two types [6]:

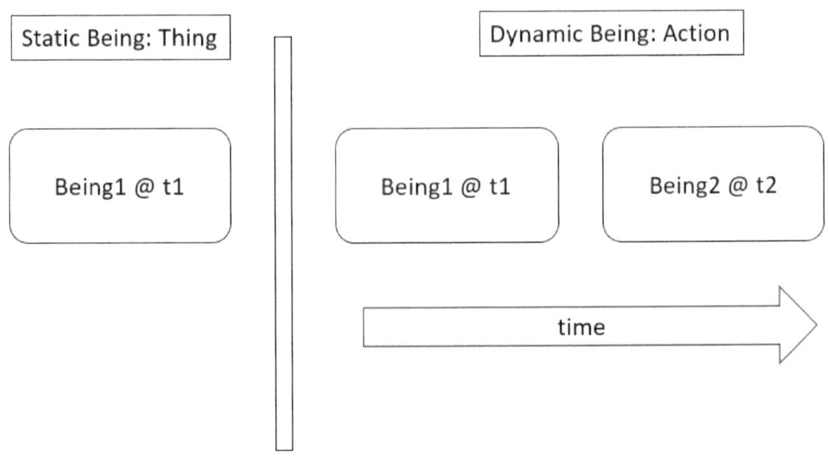

Figure 8. Static Being: Individual and Dynamic Being: Action

1. Static Being
2. Dynamic Being

A static existence is a reality which is viewed without any temporal sequence, i.e. existence at a given time. When reality appears in a temporal sequence it is dynamic and called an action [6]. The two types of being are depicted in Figure 8.

So existence/being without a temporal sequence is called substance, individual, "dravya" or the thing. In Figure 9. we see that the Genus is that which is constant across a change and the Quality is that which undergoes changes. So a thing is a collection of what is constant (genus) and that which changes (quality). A thing possesses genus and qualities.

Thing: Type of Being (Static, viewed without temporal sequence)
Action: Type of Being (Dynamic, viewed with temporal sequence)
Genus: Something which is persistent across change. And as it has existence, it falls under the category of Existence/Being. So every Genus or Jati an object possesses is Existence/Being itself. Where Existence/Being is the highest universal.

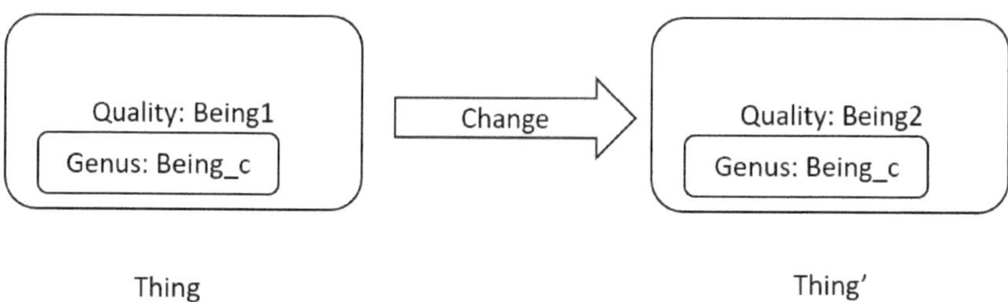

Figure 9. A Thing Changing

So Being can be classified from our discussion as shown in Figure 10.

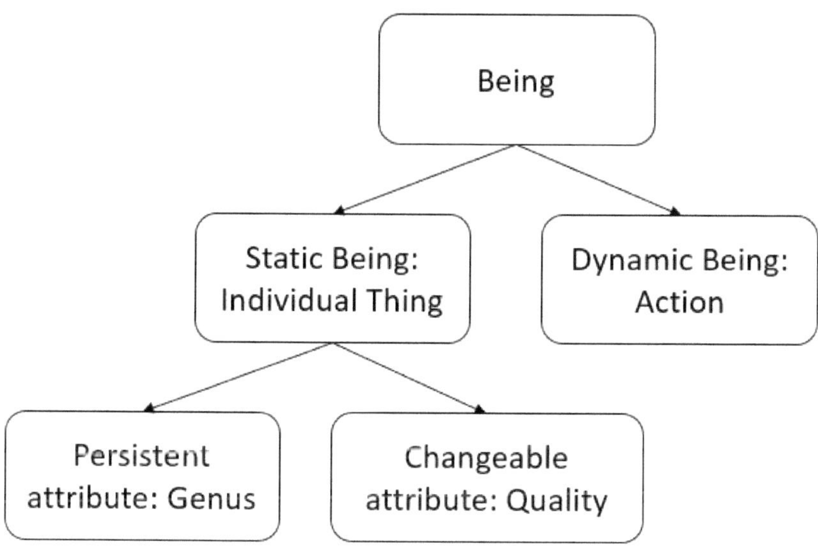

Figure 10. Types of Being / Existence

If a word refers to the genus as its meaning, it refers to something which is a Being, as Being is the highest genus and all genus fall under it. An Individual or thing has been shown to imply a static being and also as it is existent, it falls under the category of Existence. So whatever the word meaning (Thing or Genus) it would always indicate Being/Existence. Or simply put a "thing" exists and a "genus" exists so they fall under the category of existence or being, i.e. they are a type of being or existence.

So it has been proved that all words(which indicate a thing or genus) indicate a type of "Existence" or "Being". Which implies they indicate 'Existence'.

The four types of things and the four types of words:
So as discussed before the four classes of words are:
1. Genus word
2. Quality word
3. Action word
4. Proper name word (a class instance)

These four classes of words signify the four classes of things namely class, quality, action and class instance.

What is a Universal: Something which exists in all its substrata, as a result of which all of them produce a uniform awareness and are called by the same name.

So a "genus" exists, "quality" exists and an "action" exists. They have existence(a universal). These entities exist so they are a type of being/ existence or fall under the category of "being" or "existence". So we can conclude that the universal of the universals genus, quality and action is "Being" or "Existence", **henceexistence is the highest universal as all other universals come under it.**

So the conclusion is that all words (four kinds) refer to the universal "existence/being". And all words are a type of existence/being (as they belong to that class) or more precisely, the universal conveyed by a word is a type of "existence/being".

Significance of all words conveying "Existence/Being":

- This existence or being is called "satta" in Sanskrit. Hence all words convey objects which are manifestations of "satta" ("Existence/Being"). [7]
- All objects are a type of "existence" or "being". They belong to the "absolute existence" genus.

- Existence inheres in all objects as even a non-existent object exists in the mind [8]. Absolute existence pervades all objects and also all objects are a type of it. It is everything and everywhere as well.
- Existence is one only. It is a common characteristic of all entities. It is all-pervading and filling all space, very large in its extent, and it is called "*Brahman*".
- That which fills, that which swells, that which expands, that which is everywhere and in all things—That is the completeness, the fullness of Reality; and that is called "*Brahman*" in the Sanskrit language [9]. All existent things are nothing but a type of existence itself or pure existence hence everything is a type of existence. It sustains or upholds everything as without existence an object cannot be.

"Existence/Being" in ancient Indian religion:

- According to Hinduism the 'satta' (existence) in each and all is God. In Hinduism, it is referred to as Paramatma or the supreme spirit [10].
- The all-pervading 'satta' ("Being") is referred to as Vishnu in the Puranas (as it is all-pervading). Vishnu is the supreme god of Hinduism [11].
- "In the beginning, this [universe] was "Being/Existence" (Sat) alone, one only without a second". He desired, 'I shall become many and be born. He performed Tapas (austerities). having performed Tapas, He created all this (whatever we perceive). Having created it, He entered into it. Having entered it, He became the manifest and the unmanifest, the defined and undefined, the housed and the houseless, knowledge and ignorance, truth and falsehood, and all this whatsoever that exists. Therefore, it is called Existence', according to a sacred Hindu book [12].
- Existence is all existent and non-existent entities. Ahimsa and why to follow it stems from the fact, that he who injures living creatures, injures Vishnu(or God): for Vishnu(Existence) is all things [13].
- *Maharaja Prahlada* has mentioned in the *Vishnu Purana* "the whole universe is the manifestation of *Vishnu*. Search for the identity of Vishnu in all creatures. True worship of *Vishnu* consists in treating all equally" [14]. So love one, love all, as it is him only.

Notes

1. Sahitya Darpan, II.4, page 43.
2. Relationalism: A Theory of Being, Relationalism: A Theory of Being, page 19.
3. JF Staal, Contraposition in Indian Logic, Studies in Logic and the Foundations of Mathematics, Volume 44, 1966, Pages 634-649.
4. Peter M. Scharf, The Denotation of Generic Terms in Ancient Indian Philosophy: Grammar, Nyaya, and Mimamsa, Page 24.
5. Peter M. Scharf, The Denotation of Generic Terms in Ancient Indian Philosophy: Grammar, Nyaya, and Mimamsa, Page 27.
6. Karl H Potter, Encyclopedia of Indian philosophies, Volume V Pg 108.
7. Vakyapadiyam of Bhartrhari, III.1.35 commentary.
8. Vakyapadiyam of Bhartrhari, III.1.34 commentary.
9. https://www.swami-krishnananda.org/upanishad/upan_06.html.
10. Vishnu Purana 6.4.37
11. Vishnu Purana 6.4.37
12. Taittiriya Upanishad – Brahmananda Valli – 2-6.
13. Vishnu Purana 1.2.10-13
14. Vishnu Purana 1.17.90

CHAPTER THREE

The Attributes of Existence

Existence has the following Attributes:

1. The generic part of everything (unchanging part) is the existence or a manifestation of existence.
2. It is the highest universal.
3. It is everywhere and in everything. This universal is in everything, hence things exist. Everything is a type of being and existence; hence it is everywhere.
4. So existence inheres in all objects as even a non-existent object exists in the mind. Absolute existence pervades all objects and also all objects are a type of it.
5. Existence is one only. It is a common characteristic of all entities. It is all-pervading and filling all space; very large in its extent, it is called Brahman. As it pervades everything, in Sanskrit it is called "*Vishnu*".
6. It sustains or upholds everything as without existence an object cannot be.

"Existence/Being" is manifested as everything movable and un-movable in the universe its attributes are:

1. It is the manifest and unmanifest.
2. It is the defined and undefined.
3. The housed and houseless.
4. It is Knowledge and ignorance.
5. Whatever exists it is.
6. Supports the cosmic manifestation.
7. Cause of all activities.

8. It is outside and inside.
9. It is non-moving and moving.
10. It is far and near.
11. The supporter, destroyer, and creator of all beings.
12. Indestructible.
13. Formless and with form.
14. Everything and everywhere.
15. Sees through all eyes, hears through all ears, eats through all mouths, feels through all hearts, thinks through all minds, and reasons through all intellects, as he is everything
16. Has innumerable hands and legs.
17. With hands and feet everywhere, with eyes, heads, and mouths everywhere, with ears everywhere, he encompasses everything in the world.
18. Existence is fire, sun, air, stars, and the moon.
19. It is woman, it is man, it is the youth. it is the maiden too. It is the old man who totters along, leaning on the staff. It is born with his face turned everywhere.
20. It is the thundercloud, the seasons, and the oceans. It is without beginning. It is the Infinite. It is from whom all the worlds are born.
21. It possesses countless heads. All heads, all eyes, all hands, and all feet belong to "Existence". It works through all hands, eats through all mouths, sees through all eyes, hears through all ears, walks through all feet, and thinks through all minds.
22. It is the internal Ruler of the universe.
23. It is great because, as the sun it gives heat and light, as the moon it gives light, as earth food and shelter, as the oceans and rivers water as your father, mother, brother and sister love and affection.
24. The enjoyer, the enjoyed and the enjoyment.
25. Appearing as Many due to the multiplicity of its powers.
26. It is the creator, destroyer, and preserver of everything.

Notes:

1. Taittiriya Upanishad – Brahmananda Valli – 2-6.
2. Swami Sivananda, Essence of the Svetasvatara Upanishad.

CHAPTER FOUR

Conclusion

The Grand Unified Theory is that all things are a manifestation of "Existence" and are different types of "Existence". Hence all forces, particles and even time and space are a manifestation of "Existence" or the "State of existing". The performer of the action, the receiver of the action and the action are manifestations of the same "Existence".

9 798887 839080

Printed by Libri Plureos GmbH in Hamburg,
Germany